OUTSPINNING THE SPIDER

JOHN KIMBERLY MUMFORD

dition et mise en page : Culturea (Hérault, 34)
etrouvez notre catalogue sur http://culturea.fr/
ontact : infos@culturea.fr
nprimé en Allemagne par Books on Demand Gmbh
esign typographique et layout : Derek Murphy et Reedsy
BN : 9791041985388
ate de parution : février 2024
e livre :
- a été composé avec une police open source libre de droits dessinée sur Open Fondry https://open-foundry.com/

- permet de replanter un arbre. SHS Éditions soutient Plantons Pour l'Avenir, fonds de dotation pour le reboisement des forêts françaises
400 projets de reboisement de forêts soutenus depuis 2014 à travers la France https://www.plantonspourlavenir.fr/

CHAPTER I. WIRE AND MODERN LIFE.

It is the wire age.
Modern life, in all its intricate bearings, runs on wire. Wire everywhere; in the heavens above, the earth beneath and the waters under the earth. In all the legerdemain of science, which has put nature in bondage, wire is the indispensable agent.

A curious, slow, finical little trade at which the smiths of forgotten races toiled and pottered and ruined their eyesight for unnumbered thousands of years has become, within less than a century, under the spur of modern need and modern driving power, the pack-bearer of the world and the mainspring of every activity from the cradle to the grave.

Wire still makes toys and gewgaws as it always did, but it is no longer the plaything of vanity alone. Cancel wire and wire rope and their concomitant, "flat wire," from the inventory of human assets tomorrow, and the world would stop stock-still.

"WIRE AND THE COMMUTER"

This is not hyperbole. Picture yourself starting for business in the morning if there were no wire and see what the verdict would be by quitting time. Considering the vital part that wire plays in the growing and transportation of food for man and beast, it is likely you would go breakfastless after sleeping on a bed without springs or the luxury of a woven wire mattress. But that would be only the beginning of sorrow. The trolley would stand dead. Perhaps you are a commuter and journey to town by steam road. The ferry would hug its slip, and where is the railroader who in these days of congestion and short headway would dare to send a train out without the protection of the little lengths of bonding wire between the rails, that articulate the block signal system?

You could telephone the office? How and over what unless wire were used? Wireless? Without the coils and armatures that keep the instruments going or the aerials that seize the word wave in its flight, there would be no wireless.

WITHOUT WIRE—NO WIRELESS

Suppose you managed to get there. Without wire rope no insurance company would let an elevator get higher than the second story, and you couldn't signal the elevator anyway, for the annunciator operates only by an ingenious system of wires, and the control is even more complex.

You can climb the stairs, but the door key is flat wire and the shank on which the knob turns is square wire and half the lock is wire. More trouble. The buttons on your suit are flat wire; so are your garters. As for the stenographer, if she got there at all—for she is as completely wired as a telegraph system, from her hat to her shoes—the index files and office books and letter hooks and much of the other equipment of the office would fall to pieces without wire, and the machine which is her pride and the symbol of her dominion is about all wire of one kind or another, except the frame.

Distinctly, it would not be your busy day. You might spend it looking out of the window at the ships going down the river, but unhappily, the majestic liner is compact of wire, from her glistening trucks to the deepest shadows of the engine room; or airplanes soaring and swaying above the teeming town and far-stretched waterways. But an airplane lives by wire. It could neither fly nor steer nor even hold together if its frame were not strung with wire and its wings and ailerons and fuselage bound and braced and its machinery vitalized by divers forms of wire and wire strand and woven wire cord.

Far over the town and across the Jerseys you would see columns of smoke rising from busy factories—save that the mines of coal and the wells of oil are both dependent for every atom of their product on wire rope, and the lumber and metals which are the bases of industrial manufacture are in the same boat. And as for electric light—you might linger till dark but turning the switch wouldn't help, for the big subterranean cables and the multitude of littler wires that make a pathway for the current, even the dynamos with their masses of wire, they were all dead long ago.

Gas? Made of coal and oil. There would be nothing left to do but to grope hungry through dark streets and, if you could find a wireless bridge, go back to Lonelyhurst, where you would learn that without wire there is no domestic joy in this earthly tabernacle, for from cellar to roof, from the bale and rim of the coal-scuttle and the binding of the broom, from the cooking pots, the dishpan and all other culinary utensils to the baby's toys and mother's corset and hairpins and needles and safety pins and pins, it is all wire one way or another. The family would never know what time you got home, for the watches and clocks are largely wire; and there would be no possible relief in going to the club, for nobody would have a car that would run—or a cork-screw, even in the dark.

WIRE HOLDS THE WORLD TOGETHER

It is wire that has brought the world together and holds it together, and when the wire mills stop, as even they would have to do if there were no wire, modern civilization might as well be dead, and it would be. Even war would peter out. Populations might perish from hunger and probably would, but they'd have to stop killing each other except by primitive methods, for without wire, which controls the movement of ships and airplanes and submarines, and permits by telegraph and telephone the manœuvering of prodigious armies and binds the shining bodies of great guns and makes most of the instruments of precision for aiming them, war would no longer offer much chance for machine-made glory. As a guarantee of perpetual and worldwide peace no League of Nations could begin to compare with the elimination of wire from the world's catalogue of weapons.

Wire is an influential member of that family of material giants which have come into greatness within a relatively short time but which none the less weigh heavily in the destinies of mankind. It is old, too, but until a new demon of material ambition began to stir in crowding populations it had little purpose except to adorn the raiment of the great or add richness to ancient arts. People whose vision of man's past is bounded by the encyclopedia have been told times enough that Aaron's robe had gold wire threads in it, that there was wire in the pyramids, that Nineveh was beating out wire eight hundred years before the tragedy of Calvary, and that metal heads with hair of wire were found in the ruins of Herculaneum and are now again entombed in the showcases of the Portici Museum.

THE AGE-LONG USE OF WIRE

In a world chasing the present and future dollar ethnology moves slowly; the encyclopedias have not yet told that pre-Inca Peru, hiding in its tombs the secrets of a vanished civilization, has now given up garments gleaming with woven metal, which show their makers to have been past masters ages ago in the wire-beater's art, and to have spun the wire on woolen filaments in the self same way of lamination in which Paris does it for the uniforms of haughty major generals today.

And yet, down to the century when the popes were ruling from Avignon, when Rienzi was raising hob in the streets of Rome and titles of nobility were being won on the bloody fields of Crécy and Poictiers and Bannockburn, none of the many metal workers, through all the ages and in all the lands, ever had a notion he could draw metal through a die to make a wire. They hammered and hammered through the ages and sliced the filaments off as a cobbler cuts leather shoestrings—or used to. And then it was a German that did it, for the ancient records of Nuremberg and Augsberg tell of a "wire drawer" and later on one Rudolf had a wire mill at Nuremberg. The chances are that Rudolf was a capitalist and that the inventor sold him the invention for a pot of beer, and grumbled for the rest of his medieval days after the manner of his kind.

Six centuries have gone since then, and in a world of wire it is safe to say, on the strength of some inquiry, that ninety per cent of the people whose lives and well being hang on wire from one year's end to another have no more knowledge of how drawn wire is made than the Egyptian who hammered out his quota in the days of old Rameses.

THE BEGINNING OF THE WIRE AGE

England and France, quick to see what the process meant, even to the slow commerce of those times, fussed away for another three hundred years, trying to perfect methods of wire drawing to the point of independence in the trade, but it was a stern chase. "Iron wire," for all utility wire in the beginning was drawn from Swedish iron, was beginning to take up a share of the white man's burden. Gold and silver and platinum and bronze were still favored in ornamental use, but for practical purposes iron refused to be displaced. Great Britain essayed in 1750 the making of wire from steel for musical purposes, but to 1769 Broadwood was still sticking to German iron and even in 1790 was still buying wire from Pohlman in Nuremberg. So Bavaria, where first the idea of drawing metal had been hatched, was still leading the world in its craft.

Little by little, for the tide of industrial activity had barely begun to rise, new uses were found for wire. In one field after another it supplanted vegetable fibre where strength and durability were essential. As the world began to feel the Nineteenth Century surge of mechanical impulse, as life developed new facets and new needs, science sought new means of meeting them, and in the quest itself grew. Producing methods advanced with the new demands of invention. Always the wire makers spun their filaments a little finer. Men were weighing zephyrs and measuring the infinitesimal, and needed tools of increasing delicacy. Wire was the answer.

DREDGING

Electricity, so long hidden from understanding, was led captive by a wire, not by a chain—and with its development wire has found a new and increasingly important role. The ductility of metals was at last being tested to the full. Seven one-thousandths, three one-thousandths, one one-thousandth—one record after another was passed. At last, by way of curiosity, a wire was drawn that measured one four-thousandth of an inch in diameter—twelve times finer than the hair on your head. The spider, so long counted a master workman, had been undone.

The wire age was arriving—big wires to carry the world's heavy loads; fine wire to solve its molecular problems. The day of the hammer was done.

CHAPTER II. THE PIONEER.

Since Columbus the centuries have been gathering speed. At first it came slowly, for the need was not yet. Today a thought is born and tomorrow it is a giant, parting seas and moving mountains. The waste of yesterday is turned into the raw material of new manufacture, with its million wheels moving faster and faster. But back of it all, inevitably and eternally, is a busy human brain and unsatisfied energy.

Wire rope had lingered, waiting for civilization's loads to grow. The artisans of old had woven cut wires together to make the trinkets of their time, little dreaming of the might that lay hidden in the fibres of the iron, and their world went on hoisting stone for its pyramids by prodigious multiplication of garlic-fed man-power. It seems strange to the high-speed mind of today that five hundred years could have passed, after the drawing of wire was invented, before necessity put it into the mind of a wire-drawer that with wire, as with other things, strength lay in union. And yet the human race had been making rope since the morning stars sang together.

In 1831, when France was picking herself up from the dirt and disorder of another revolution and the German princes were strangling in the universities the growing call for "liberty and union," young men of brains and ambition began to leave the German states for America, where there was free air and elbow room.

JOHN A. ROEBLING COMES TO AMERICA

In a company of such, John A. Roebling journeyed from Muhlhausen in Saxony, and took up a tract of land in western Pennsylvania. He carried a degree of civil engineer from the Royal University in Berlin; but there were "back-to-the-landers" even in those days, and he set about farming in the thrifty German way, founding for nucleus a little town which at first was named Germania, but afterward came to be called Saxonburg.

Fate seems to have ordained that Roebling's engineering skill should not remain fettered to a Pennsylvania plow handle. The system of canals and portages which afterward evolved and merged and built itself into the Pennsylvania Railroad was digging its ditches and dams and building haulways through the obstinate distances of that hard-ribbed state, past the hopeful hamlet of Saxonburg and fatefully under the eyes of the young German engineer. The result was never in doubt. He abandoned the plow to his compatriots and plunged into the problems of construction, where he belonged.

HAULING CANAL BOATS UP THE PORTAGE RAILWAY

The skeptic who scoffs at fatalism will find it difficult to explain why the particular engineering work that was brought to Roebling's door should involve the weary hauling of the Pennsylvania Canal's boats up the Portage Railway, which Bertrand, one of Napoleon's generals, had built to overcome the Pennsylvania ridges; or why, just as the bulk and clumsiness and inefficiency of the huge hemp cables were eating into his active mind, a casual paper from Germany should convey the fact that some fellow in Freiburg in Saxony—where wire drawing had birth—had made a strong rope by twisting wires together.

What man had done man could do. If there was a place to test the efficacy of wire rope with its increased strength and diminishing size, it was the Portage Railway. So the Saxonburg wheatfield was turned into a ropewalk. Ceres made way for Vulcan. The neighbors, as soon as material could be shipped in from the Falls of the Beaver River, where wire drawing was done, found themselves under young Roebling's direction twisting wires, with rude appliances for torsion, into a fabric which had never been made or seen or probably heard of in America before, but which was destined, in a comparatively short time, to change the face of industry.

WIRE ROPE PROVES ITS PULLING POWER

It is easy to imagine the caustic comments of the Pennsylvania countryside, and the forebodings with which the pioneer installed his cables on what was then a conspicuous engineering labor. But it worked. Engineering audacity, plus scientific skill and native faculty for doing things, solved the problem of the Portage, but it did far more than that. The fame of it was soon broadcast and the orders for wire rope came flooding from all that fast opening country. Roebling had found his job. Destiny had him by the collar and he bade farming good-bye.

HOISTING A BATTLESHIP TOWER WITH WIRE ROPE

It was in 1840 that the first Roebling rope was finished. Eight years later, the year when the revolution burst forth in the Teutonic empires, he moved his plant and its business to Trenton, and began forthwith to build the foremost wire rope factory in the world.

Nothing can be more amusing or reveal more clearly what brains and energy have been able to accomplish in the arena of American opportunity than to contrast the picture of the first Roebling factory in Trenton, which suggests the rudest of farmsteads, with the sky-piercing chimneys and the mile or more of many-windowed brick buildings in and around the Jersey capital today, where the Roebling work is done.

The three big factory groups which have grown from the shabby little buildings of 1848 are the fruit of intelligence and ceaseless endeavor, but they are reared primarily on a basis of manufacturing honor, and ruled by the general thesis that forever and ever quality comes before price. This means keeping faith with the structural iron worker, swinging pigmy-small five hundred feet above the din of the city streets; with the sailor, the miner, the rigger; with the hurrying multitude that packs the elevators in tall buildings, and with the aviator, to whom a breaking wire may spell death.

That is the reason the Roebling Company has outgrown the limits of Trenton in the last decade and a half and with its overflow founded a city of its own; that is the reason why Roebling has almost got into the Thesaurus as a synonym for wire in every civilized language under the sun.

It is wire, from the huge three-inch cable that pulls the loads of mountain haulways or moves the thousand cars of a city transit system, down to the gossamer that jingles the bell in the telephone or the infinitesimal hair that in the eyepiece of a telescope helps the astronomer to mark the movement of a distant world. There is hardly a thing in the nature of wire, round, flat or irregular, that the Roeblings do not manufacture or have not at some time manufactured, whether for the world's standard uses or the numberless special purposes hidden in inventive minds.

A TWELVE MILLION POUND DEVELOPMENT FROM A FIFTY POUND BEGINNING

"I've come to see," said an old man at the Roebling offices one day, "if you'd go to the trouble on a very small order to find out just what composition I need in a wire for a patent I've got."

And they did. It took the chemists and the experts some time to work out the problem of resistances, and the old man ordered fifty pounds. The next year he ordered a hundred more. There was no profit in it, but they made it and looked pleasant. They were specialists in wire and they were simply keeping faith with their job.

The following year the visitor called again. "I don't want any more of that wire," he grinned, "I've sold my patent to So-and-So," naming one of the biggest manufacturing concerns in the world, "but I want to see some royalties and I made it a condition of the sale that they order this wire from you on the formula that I got."

In a recent 12 months period Roeblings fabricated more than 5,000,000 pounds of that wire.

If it's wire, the Roeblings make it. All that was in the mind of the man who seventy years ago was twisting the first rope in Saxonburg. He was more than an engineer; he was a sane and far-seeing mind in business. As soon as possible after establishing the factory in Trenton he added a mill for the manufacture of his own wire. It gave him a product that he knew from the pig iron up, and it saved a profit, besides extending to a marked degree the scope of the business. He knew, when he put the cable on the Portage haulway in 1840, that the mission of wire, in the world that was then making, would be boundless, and from the very start he was the explorer in new fields for wire, a builder, a seeker for problems that wire might solve, archapostle of the power of wire, in one form or another, to do the heaviest labor of mankind.

Wire rope, spreading its field of utility ever wider and wider, carried with ease and safety loads that had broken the back of hemp; it took the place of solid steel in numerous phases of construction, and when its adaptability was proven new tasks were devised for it. Wire rope was the forerunner of "Safety First." It cancelled large burdens of expense; it set a new record in facility of construction.

AMERICA'S FIRST WIRE CABLEWAY

Persistently militant, from the day of his first achievement, in the promotion of wire rope, John A. Roebling was the first engineer to introduce into America the novelty of a wire cableway, which with an ingenious carriage he employed to transport across a river the materials he needed in the construction of a bridge. This method of haulage, over streams and gorges, down from high mountains to cars or boats in the valley below, up from the deep-sunken beds of rich placers—everywhere and in all sorts of places where Nature seemed to have set up impassable defense against those who would take away her treasures—came forthwith into widespread use, and is among the handy tools of engineers throughout the world today. The Roebling Company established these cableways in many countries. It had in operation around the globe no less than twenty different types, including log rigs and gravity planes for mountain railways, and the demand for wire rope was increased thereby a thousand fold before the new century had come in.

ROEBLING TURNS HIS ATTENTION TO BRIDGES

The age of wire was marching rapidly, but John A. Roebling had set a distant mark. In the mountains of Peru, India and other lands for ages the natives have made use of bridges made of vines, to cross appalling chasms. As time went on and arts progressed the principle was applied through the agency of hemp ropes and chains, and men of small imagination thought that in these the limit had been attained. But Roebling's faith was as the faith of the Moslem in the Prophet. He believed that in wire the solution of all the pesky problems of bridge-building had been found. In a small way the thing was obvious, but his ambition never stopped there. He believed, and had believed ever since he made the first rope, that a major bridge made up of wires of scrupulously high quality, constructed with rigorous regard for scientific tenets, would carry with ease and indefinitely any reasonable traffic that might be imposed on it.

Famous engineers said he was a visionary and a hobbyist; still with force and tenacity he urged his contention until at last the engineering world was compelled to give heed to him. In the face of such opposition, and in view of the centuries that had dragged by before wires were twisted into rope, it is remarkable that so soon after his initial experiment he should have worked out in practical entirety the plan of bridge construction which came to its climax in the spanning of the East River.

Between 1840, when he made his first rope, and 1844, he had not only perfected his theory of wire bridges but in spite of furious opposition had built one as an aqueduct for the old Pennsylvania Canal, the basins of which were at Pittsburg. This was followed by four more suspension aqueducts for the Delaware and Hudson Canal Co. Having espoused a theory he let no grass grow under his feet. He cast about vigorously for bridges to build. He found an opening in Cincinnati.

THE OHIO RIVER BRIDGE AT CINCINNATI

River traffic along the Ohio, in the forties, was still a big factor in business but was contesting tooth and nail the advance of the railways, and fought bitterly against the right of the invaders to build bridges over the waterways. The steamboat men said bridge piers would be a peril to navigation, but the cities of Cincinnati and Covington, facing each other across the river, cried for the bridge. The rivermen were on top in 1846 when Roebling came along, fresh from the building of the wire bridge in Pennsylvania and with his head full of wire bridges, and offered to throw a wire span across the Ohio with a length of 1057 feet and a floor height above the water of 103 feet.

LOGGING – HANDLING BIG FELLOWS WITH WIRE ROPE

For just ten years the steamboat faction staved it off. It was not begun till 1856, just after the Niagara Bridge was opened. The panic of 1857 and then the Civil War kept the project at a standstill until 1863. On Easter Day in 1867 the bridge was opened. Colonel Washington A. Roebling, son of the pioneer, was the first to cross on its cable. In the meantime John A. Roebling had completed not alone the Niagara Bridge, but the Alleghany Bridge over the Alleghany River at Pittsburg. The last named differed from the Niagara, Ohio and later East River bridges in that it had several piers in the streamway, after the manner of the old type structures, but in principle it conformed to the plan which had been in his mind from the beginning. His son, Washington, was his only assistant.

BRIDGING NIAGARA GORGE

In all the world, perhaps, no place could have been found where the building of a simon pure "Suspension Bridge" would have been a more spectacular accomplishment than over Niagara Gorge, with the Falls thundering a little way upstream, and the waters lashing and fuming underneath; no place where its slender beauty could have had such stern and impressive background. The idea of carrying railroad trains over that turmoil of waters on a web apparently so frail, evoked a storm of protest from well-nigh all the foremost engineers of the time. But Roebling was a practical man as well as a stubborn one. After all, he was dealing with rock and wire and he knew what they would do. He built the bridge, the first of its kind to carry railroad traffic. All the world of that day knew, but most of it now has forgotten, how he flew a kite across the gorge to get his first wire over, and from that built up his cables. On March 16, 1855, the first train passed over it. With one remodeling it continued to carry increasingly heavy loads until nearly half a century later it was replaced by a larger structure, better calculated to bear the burden of modern equipment.

THE "SUSPENSION BRIDGE" PROVES ITSELF

"Suspension Bridge" not alone proved itself in point of service, but it demonstrated the soundness of Mr. Roebling's claims for the wire structure. The Ohio structure, which followed, outdid Suspension Bridges in length of span; in economy of material, in simplicity and charm of outline it clearly foreshadowed the still greater work, the designing of which was to be the crowning accomplishment of his life. He was working with a practiced hand now.

The doubts, if he ever had any, were behind him. Behind him, also, was a producing plant tuned to turn out at speed the materials he needed, with certainty of their quality.

He had proved that the making of big bridges with wire was feasible, and that it was simple, as most great things are after they have been done. There were only three basic parts to a suspension bridge after all—towers, cables and anchorage. Suspending the roadway, which to the average man seems the vital part of the creation, is, from the engineering standpoint, only an accessory work. John A.

Roebling had concentrated his life's effort, not on mere methods of commercial production, but rather on the proving of his contentions. He needed the right kind of wire rope to prove them, so like a wise man he made it himself.

He came to the summit of his achievement with the acceptance of his plans for the building of the Brooklyn Bridge, and then, his faith vindicated, his theory, which he had fought so hard to sustain, endorsed by boards of noted engineers and acclaimed by the public, starting out on the realization of his long dream—the building of the Eighth Wonder of the World, a comparatively slight accident, the bungled docking of a ferryboat, which crushed his foot and brought on tetanus, put out the steady candle of his life.

It was the very whimsy of fate. His work was done. He had created, out of imagination and energy, the finished designs for a wonder fabric, ready for the labor of an intenser age. He did not live to see the spider structures hung like wisps of gossamer above the restless waterways of New York, but his name is woven into the very steel of them.

CHAPTER III. THE BROOKLYN BRIDGE.

Early in the fifties, when the Niagara accomplishment was more or less the talk of two continents and communication under seas by cable had helped to emphasize the possibilities of wire, John A. Roebling, protagonist of the wire bridge idea, advanced a proposal to connect New York and Long Island by a suspension bridge and release the people of Brooklyn from a segregation which they had made a somewhat futile pretense of enjoying. Habit dies hard. The crust of custom becomes strangely indurated with long exposure, and Brooklyn residents had fought the East River in profitable, if archaic, ferryboats too long to be lured lightly into any liaison with iconoclastic Manhattan by way of a wire bridge.

Roebling waited another decade, but he hustled while he waited. The Brooklynites continued to make their uncertain ways across the river in times of storm and tide and ice as the Lord gave them strength, and the sacred ferryboats still paid dividends. The vicious winter of 1866-7, coldest, bitterest, longest the cities have ever known, wrung forth at last a cry for relief. They could wrap themselves up against the weather, but no weight of woolens could turn the shafts of ridicule. It was grand ammunition for the advocates of the bridge, when people traveling by train from Albany actually reached New York sooner than did the man who did business in New York, and left his domicilium in Brooklyn at the same hour.

And besides, the Roebling cap had another feather in it now, in the completion of the Ohio Bridge. He was building wire bridges everywhere, and it began to look as though there was some body of truth in the Western contention that New York was the most provincial city in America, for all its self-approval.

At one of the many hearings that were held on the bridge question a famous engineer who favored the wire type was asked what reason he had for believing it would do the work.

"I believe it," he replied, "because Roebling says so."

THE INITIAL CHARTER GRANTED

The demand for the bridge rose to a clamor. In the month of May, 1867, the initial charter was granted, and Mr. Roebling was appointed engineer. Three months afterward he submitted his report and estimates, which were examined and approved by a commission of engineers from the United States War Department. Then he set about preparation for the task.

THE DEATH OF JOHN A. ROEBLING

It was while fixing the location for the Brooklyn tower that he met with the accident that caused his death. But his work had been well done, and his son and associate, Col. Washington A. Roebling, took up without delay the execution of the plan he had helped to create.

If the older Roebling encountered obstacles in bringing his great idea to the point of acceptance, the pathway of his successor, called without warning to take over responsibility for the greatest engineering labor of the age, was not strewn with roses.

THE WORK OF CONSTRUCTION BEGINS

WIRE ROPE IN THE QUARRY

It was in the summer of 1869 that John A. Roebling died. The second day of January, 1870, saw the actual work of construction begun, when laborers started to clear away to prepare for the foundations of the Brooklyn tower. From that day forward, through a baker's dozen of years, there was no rest, though there was plenty of interruption. Until the job was ended Washington A. Roebling simply lived the Brooklyn Bridge. It was a colossal job, punctuated with changes and problems and complications, but it went forward. The landmarks of a bygone age, old houses of historic memory on the water fronts of both cities, vanished silently and where they had been, by and by there grew piles of masonry to form the approaches. From the huge caissons over against either shore rose the towers, tall and grim, which were to carry the cables. In due time they stood complete, with their broad bases welded to the rock by an ingenious bond of stone and concrete in the river's bed, and their crests nearly three hundred feet above the top of the tide. A hundred and nineteen feet—and three inches, to be precise—above the water opened the two tall arches in each tower, stretching upward one hundred and seventeen feet in air. It was through these the bridge proper was to pass, with its gangways for horse and foot and railway traffic.

COULD THOSE SLENDER TOWERS CARRY THE GREAT LOAD?

The hurrying people of New York and Brooklyn watched the thing grow and wondered fearfully whether the slender towers would stand the strain. In Harper's Magazine for May, 1883, now itself yellowed by age, is an exhaustive article concerning the Brooklyn Bridge, in which one is told at length and with an engineer's exactness, the steps by which the achievement was brought, after thirteen laborious years, to proud completion.

Even to the curious layman the details are no longer of insistent interest. One thing is emphasized, however, which well as we know it now can never cease to hold the mind in a certain wonder—that all the weight and solidity and massiveness are in the towers, the foundations and the long expanses of stone work, which stretching inland nearly a thousand feet, serve to guard and strengthen the anchorage for the cables which are the working force. The rest is wire, for the most part; wire, slender by contrast and against the background of the sky, but endowed with great strength by care and skill in fabrication. John A. Roebling and his son had staked their name and their future on the strength and quality of Roebling wire.

In that long ago story of the Brooklyn Bridge, there is written the lesson that clear thinking and courage and perseverance can accomplish the seemingly impossible. What traveler over those high-hung roadways ever stops to ask himself how those great round cables, stretched in long, inverted arches above the surge of the river traffic, were ever put in place? They are today simply a part of the stage setting of a busy life, like the river itself.

HOW THE GREAT CABLES WERE MADE

Each of these cables consists of nineteen strands of about two hundred and seventy-eight No. 8 B. W. G. wires each, and each wire is continuous in its strand, like the yarns in a skein, traveling eternally to Brooklyn and back, up over the top of one tower, down in a long curve above the tideway, up to the other tower and down again, to be gripped and carried by links, like a chain, down to the everlasting clutch of the rock and concrete-bound anchorage. Each skein is a million feet long—nearly two hundred miles—and still men talk of "Oriental patience."

There is no twist in these ponderous cables, as there is in a wire rope. Every reach of wire lies flat and separate, and when all were in place they were laboriously bound together, first the strands, then when all the strands were up, the whole fabric, into cylindrical form. There are other strange things about these cables; one is that they make practically no strain on the towers save to sustain their weight. Another is that the long storm cables that radiate downward from the top of the towers to the bridge floor, for a space of four hundred feet inside and outside each tower, are themselves calculated to sustain, if need be, the imposed weight for that distance. So that the margin of safety in this seeming web-like structure is far in excess of what timid imaginations have pictured. That was a cardinal feature in all John A. Roebling's plans. He left a safety margin many times greater than the load. It has been an open secret for years that the Brooklyn Bridge has been unwisely taxed, but he knew it would be.

STRINGING THE CABLES ACROSS THE EAST RIVER

Before the cables were in place, New York and Brooklyn stared up at the river-wide space between the bare towers and wondered by what wizardry a bridge could ever be swung across it. The beginning was simple—as simple and prosaic in a way as the hitching of a horse—in principle. It began with wire rope. A scow with a coil of three-quarter inch rope was moored alongside the Brooklyn tower, and the end of the coil was hoisted up the face of the masonwork, passed down on the land side and then carried back.

HELPING TO RELIEVE THE FREIGHT CAR SHORTAGE BY QUICK LOADING

Next, suspending the river traffic for the necessary time, the scow was towed across the river, paying out as she went, and the rope carried over the New York tower, then wound on a huge drum till it hung high above the river and clear of the tallest topgallant. A second rope was run in the same manner and the two were joined around huge driving wheels or pulleys at each end. An endless belt or "traveler," revolving by steam power, now stretched from city to city, and on a day in August, that lives yet in the memory of every man who was there, E. F. Farrington, the master mechanic of the project, who was a veteran of Niagara and the Ohio Bridge, set out to show the workmen, who on this slender aerial were to begin the long labor of hanging the cable, that it was easy if you only thought so. In a "bosun's chair" he shot out from the top of the Brooklyn tower, down the long sag in the traveler and up to the New York side, while a million people craned their necks from the streets and docks and housetops and boats along the river, and swallowed hard at their hearts.

The bands played, the cannon tore the air, the multitudes yelled themselves hoarse, the steam whistles of the harbor shrieked to the sky the tidings that, though nobody then understood it, "Greater New York" was on the way.

This was six years and a half from the time when Washington A. Roebling had begun the work of construction. Seven other years followed, years full of troubled effort, of planning and replanning and replanning, of battling with the twin devils of Contraction and Expansion. The tensions all had to be secured in absolutely uniform weather. A determination made when the sun was shining on one part of the bridge and not on another might have thrown the whole calculation awry. Sun and wind played pranks with the work in the summer and in the winter snow and ice coated the wires and running gear so that work was often impossible. Deflection varied a third of an inch for every degree of temperature.

"In short," says the writer of that time, "the ponderous thing, while neither small nor agile, has a trick in common with the minute and lively insect which when you put your finger on him isn't there."

THE FABRIC GROWS TOWARD COMPLETION

But in due time the great cables were in place, and bound. Then the suspender bands were set, from which suspender cables hung to hold the frame of the roadway. And so the fabric grew toward completion, hung practically in two sections, which all the world nowadays doesn't know, with an expansion joint connecting them in the middle to absorb the expansion and contraction of the metal. Even the rails at this section are split in half lengthwise, to permit them to slide back and forth with the changes in temperature.

There were accidents and drawbacks and political complications, as there are always bound to be in public works; there were believer and unbeliever, booster and knocker, as now, but the work went on to its completion and in 1883 the day of realization came. Wire was king. Doubters and malcontents murmured for a time, but little by little subsided. The opening of the bridge was one of those memorable days of which New York has had so many in her brief history, a day when President and Governor and many lesser dignitaries, who have now passed from the stage, strutted their little hour to hail the passing of a milestone, and there were "fireworks in the evening."

THE BEGINNING OF A NEW ERA

A new era had now definitely begun. There was a recognized agent in the world strong enough, with engineering guidance, to shoulder its most staggering burdens, and the name of Roebling began to weave itself in letters of wire through the whole web of modern industry. Thirty-seven years have come and gone since the Brooklyn Bridge was finished and thrown open to the swarming people. Even when they saw they wouldn't believe it; many of them mounted to its span with their hearts in their mouths. There had been a world of carping and prophecy of disaster. A public that clutched at novelty as an addict does for stimulant could not assimilate the idea that there could be safety in wire where such enormous weight was laid upon it. Its frailty of appearance fooled them. For years after the Bridge had taken up its load and was carrying without protest or misbehavior the traffic of two cities, there came periodical alarms regarding the discovery of strange faults in construction, or disintegration of the wires caused by vibration. It was the one dependable theme for the alarmist and sensational writer.

But the proof was in the using. The slender span has stood the test of time and tide and wind and wear, and stood them all so well that it has fixed for a century at least the type of the super-bridge.

TWO MORE BRIDGES TO BROOKLYN

Wire bridges have become a familiar thing in the lives of cities. Two more have come to give the crowding population of New York freeway over the East River, as the city's life has spread northward. For the Williamsburg the Roebling firm furnished the wire and installed the cables. In the Manhattan Bridge it had no part save the making of the wire, not a trivial task, since in the cables alone there are 12,000,000 pounds.

These bridges are bigger than the Brooklyn Bridge with which the troublesome river was first overcome, but it will be many a day before the glamour that surrounded the earlier creation will have worn away, or people the world over cease to speak of it with wonder and a certain measure of awe. Anybody, perhaps, can build a wire bridge now; perhaps, too, somebody some day can build one with more of simple grace and slender beauty, but it is certain nobody ever has.

CHAPTER IV. WHERE WIRE IS MADE.

To measure the growth of wire, with its many forms and composites, during the last forty years would be to trace in detail not alone the progress of science, invention and mechanical industry, but the myriad conceits that have come ostensibly to facilitate the process of living. In the search for new comforts, for means of avoiding physical exertion, the world has been littered with novelties, and most of them depend on wire. Personal life as well as commerce and industry is interlaced with wire. With the opening of new countries, the increase of populations, the flocking of outland people to the cities and the consequent lack of farm labor, ingenuity has been more heavily taxed to find the quick and easy way of doing the world's work and keeping food in its mouths. Wire, so adaptable to the heaviest as well as the lightest tasks, has labored from year to year under an increasing demand.

It is not surprising therefore that a company which in such an impressive way had fixed itself in recognition as the first exponent of wire's usefulness should have grown in this period from modest commercial stature to a high place in its field and to the enjoyment of large production.

THE GROWTH OF THE ROEBLING BUSINESS

When the sons of John A. Roebling took up control of the business he had established, about one hundred men were employed and the product of their industry approximated $250,000 annually. Just before the beginning of the war more than eight thousand employees were engaged in the manufacture of Roebling products and the value of the output ran far into the millions. The factory which was so meagre and so humble in 1848 has spread its buildings not only over the surrounding acres, but across what were then neighboring farm lands until, constrained not alone by the pyramiding demand for its products but by the soaring values of the city that had grown up around it, and of which it had been in some measure the creator, it went pioneering again, sixteen years ago, down the Delaware, and established a new nucleus, which will suffice for a long period to come.

With the erection of the cables for the Williamsburg Bridge, the Roebling firm withdrew from the competitive field of engineering contracts and concentrated all its energies in the perfection of its product—wire.

In view of the more distinctly industrial character of the Roebling enterprise under the later dispensation, it is of interest that the varied activities of John A. Roebling, as a scientist, a master of materials and a peculiarly astute mind in affairs, have been carried on severally among his sons and grandsons. Colonel Washington A. Roebling, the president of the company, who executed the plans for the Brooklyn Bridge, is an engineer of well-known ability. His intimate contact with all the affairs of the company during such a long period of development, his kindly and generous support to constructive achievements, has been a source of pride and invaluable assistance to the younger generation of the Roebling fraternity. His two brothers, Charles and Ferdinand, now dead, were both intensely active during their lives. Charles G. Roebling's talents as a builder of plants and machinery and an unusual gift of turning out a product of the highest excellence, were, in a large measure, the cornerstone for the tremendous success of the Roebling Company. It was during the period of his direction that the manufacturing capacity grew so rapidly.

The simultaneous expansion of the commercial field was the life work of the other brother, Ferdinand W. Roebling, who carried the Roebling products to all corners of the globe. A clear and far vision, an uncanny ability to go straight to the point and a keen knowledge of human nature, were a few of the strong traits of his mentality. Under his control of financial and ethical matters the John A. Roebling's Sons Company established a worldwide and enviable reputation for stability and fair dealing.

Ferdinand, although an indulgent father, brought up his two sons, Karl and Ferdinand, Jr., in the old-fashioned way. They were taught from early boyhood that theirs would be no bed of roses, that manhood was an estate where responsibility must be accepted and assumed, and with this teaching ringing in their ears the mantle of the presidency of the company fell upon Karl G. Roebling, and the secretaryship and treasurership upon the shoulders of Ferdinand W. Roebling, Jr.

Both sons upon leaving college were given a rigid training in all branches of the business and early in their careers exhibited the executive ability and keen business foresight which their father had in so large a measure developed. Karl's talents lay principally in the gift he had of drawing from his associates their whole-hearted fidelity and devotion to the cause of the Roebling prestige. His death at the early age of forty-eight was a shock to the industry, and a great personal loss to those associated with him in the conduct of the business.

While all of the Roeblings have possessed, in a great degree, the qualities of leadership, yet they have always recognized the necessity of surrounding themselves with a strong organization capable of carrying on this great industry after they had ceased their earthly activities.

It was particularly under the regime of Karl Roebling that the strong foundation was laid for the present powerful organization—each department highly specialized and in charge of experienced well-trained heads, ably aided by a corps of competent assistants, all functioning smoothly like a well-balanced machine. Karl left this as his heritage to the business. He never did things by halves. His working day was long and intense, but to one so constituted it could not be otherwise. During the world war and its aftermath the added responsibilities he so cheerfully assumed, contributed largely toward bringing to an end a life full of early accomplishments.

Ferdinand W. Roebling, Jr., the remaining son, now vice-president and treasurer, is an able engineer. His early training with the company was entirely in the manufacturing and engineering side of the business. In more recent years, however, he has devoted his attention to its financial affairs. His close contact with his father and brother, his thorough knowledge of the company's policies, have well fitted him to sustain the Roebling name and all it represents in the business world.

THE TRENTON PLANT

The main or first plant of the company centers around the site of the original buildings. Its structures, yards and tracks cover more than thirty-five acres of ground about a mile from the center of the city. The Delaware and Raritan Canal and the Trenton Division of the Pennsylvania Railroad pass along its western boundary and directly before the door of the offices. The office building was erected in 1857 by John A. Roebling as a residence and later, as manufacture crowded in around, it was given over to business uses. The spur tracks of the Pennsylvania traverse the company enclosure.

Nearest to the office building are some of the structures that Mr. Roebling built in the first periods of business expansion, among them the old rope shop, where by methods of his own devising he strove to meet the growing demands for rope. Some of the machinery he built is still in service in production of standard lines, showing how swiftly and how far, from crude beginnings, his active mind advanced along the road to better production, and how efficient management can prolong the life of a mechanism that is honestly built in the beginning.

THE BUCKTHORN AND KINKORA PLANTS

The second or Buckthorn plant lies half a mile farther to the south, also facing the railroad and the canal.

The third, which was christened Kinkora, after a neighboring village on the railroad, but is now Roebling, with a station of its own, is ten miles farther down the Delaware. All told, there are probably a hundred buildings in the three plants, many of them of immense size and manufacturing capacity.

From the wide diversity of its products, the men in the Roebling establishment have come to refer to it as a department store. The problem therefore of distributing its operations and keeping track of its large volume of moving stock and its equipment is a substantial one. While in some lines there is activity partitioned among all three plants, in the main the various divisions of labor are well concentrated. For the most part the

Upper Works, though a considerable quantity of wire is made there, is devoted to what is termed "finished product." In the same manner the Buckthorn plant, while turning out some rope in small sizes, specializes in all forms of insulation and the manufacture of lead-cased cables.

THE KINKORA PLANT AT ROEBLING

The Kinkora or Roebling establishment, carrying the production of the subsidiary New Jersey Wire Cloth Company, making wire netting, window screens and other forms of wire cloth, is given over most largely to the making of steel wire and the fundamental work of wire and steel production. With the company's large acreage at this location, its townsite and the facility of river transportation as well as rail, with unlimited water, of which this plant uses more than is pumped by the city of Trenton itself, the situation offers large opportunity for expansion and profitable centralization of operation. At the present time, while shipments of wire are made direct from Roebling to manufacturers who use it in production of their own commodities, by far the greater part of the output goes to the other plants to be finished into rope and specialties.

MAKING A CROSSING BY CABLEWAY

Inside the tall palings that enclose the great mill buildings at Roebling, there is an open space, broad and long as a drill ground, threaded by spur tracks and heaped endlessly with stacks of pig iron and steel-making materials. It seems as though some giant had dumped there the salvage of a hundred battlefields. It lies there sadly rusting under the weather, waiting the moment when the mills shall stretch forth hands and hurry it in, rush it like a neophyte through the fierce initiation of heat and chemistry, and having changed the very fibre of it by strange processes, send it singing forth, shining in great coils, twisted into cords and cables small and great, bare or insulated, bronzed or coppered, galvanized or enameled, huge and bulky or spun to hair-like tenuousness, to do its work in a busy world.

MAKING WIRE STEEL

Of course, the making of steel is no new story, but this is wire steel—the high carbon, the tough, the sinewy, the resilient, that must carry in itself as it moves along through these interminable buildings the analytically measured proportions of this or that, which fit it to bear up the traffic of a giant bridge or convey a whisper of telephonic sound or register split seconds in an Elgin timepiece. It is "pig," and ore and "scrap," but just what kind and just how much of "scrap" and ore and "pig," these are subtle questions. It costs a lot of time and money sometimes to answer them.

When the thirty-five hundred and odd degrees of heat in the long rows of open hearth furnaces have brought this stubborn mixture to bubbling and seething like a busy kettle of soup—a workman adding a little manganese or other ingredient to the broth now and then, grimy men with long handled steel dippers take out a few thimblefuls from time to time and hurry the sample away to the chemist, who, like a chef, tests the quality and prescribes the seasoning. By and by it is run off, from an opening in the bottom of the furnace into a huge caldron they call a "ladle." A fifty-ton crane conveys it down the long, shadowy building, to halt above a group of tall moulds. A wizard up in the gloom under the roof moves it from mould to mould, a few inches at a time, while the liquid steel is drawn from the bottom into one after another. The moulds are left to cool.

BLOOMS

Its history is now begun. It is an ingot—many ingots—and when removed from the mould is loaded on steel cars and borne away on its journey. When in due course the ingot comes to the "blooming mill" it is fourteen inches thick each way and five feet long. Heated again, it is marched up on a steel rollway, also controlled by a "man higher up," and into the hungry jaws of a machine that, after a series of swallowings, disgorges it at last, shrunken in sheer humility to a diameter of four inches and with a very long face—some forty-eight feet to be exact. And no wonder. In the process it has been kneaded into a dozen different phases of flatness and squareness, and put in a way to profit by the everlasting squeezing and stretching it is to undergo. Now it is a bloom.

BILLETS

Again it is passed on, and from some subterranean blackness you see it rushed out and up to a sort of guillotine that first cuts off the flawy ends, where the impurities accumulated in its ingot state, and sends them to the "scrap" heap, then lops the bloom as a man saws firewood, but a great deal faster, into billets varying from one to four hundred pounds in weight. They are "billets" now, and at last are counted the raw material of wire, even after such an inferno of cooking.

A steel loader gathers them up, carries them away in bunches and, by a trick of wire pulling, deposits them on other cars in rows as regular as the pickets on an old fashioned fence.

THROUGH THE ROLLING MILL

Along with the copper billets they are stacked in thousands and thousands of tons in the stockyard outside the doors of the rolling mill, each in its group according to physical and chemical character, waiting the next purgatory of change. One pile is marked for one mission, one for another, ranging through all the uses wire can be put to. These piles are forever vanishing, forever being replaced, as the wide world calls for wire. They disappear into the darkness of the mill and they are never billets again.

Marshaled on cars and jammed by hydraulic force into big reheating furnaces like a Brobdignagian bakery, fired with fuel gas, they come out glowing again and start on the next stage of reduction. The passage through the rolling mill is a short life and a merry one. If they were kneaded in the blooming mill it was a mild experience. Here they are mauled and manhandled and masticated by swift, continuous and looping mills that are born with a huge appetite for the largest billets, and make rods of great length. Down they go, under the gripping of relentless fingers that squeeze them first square, then oval, then square again, and pass them on, always smaller, toward the journey's end. Sometimes it's half an inch, sometimes more, according to the needs of trade.

THE MILE A MINUTE JOURNEY INTO WIRE

Wire goes the whole distance, whisking along through the murky, half dark mill, up and down at a mile a minute, like flaming serpents flirting fiery tails, as the men, armed with tongs, seize and whip them from one pair of rolls to another. In they go, around the grooved repeater and out again to be grabbed with a motion swift as the dash of a pickerel, and thrust once more into the next set of rolls. Always the lightning speed and always the long tail, red hot and smaller than before, and longer, playing "snap the whip" down the steel grooves to the bottom of the "pit," then straight away up the incline, a flash of fire in the darkness, and on from roll to roll. The men who handle these rods hold their ticklish posts only twenty or thirty minutes at a time. A straight eight hour day, if a man came through it alive, would send him to an asylum with a conviction that he was great grandson to Medusa. At the finishing pass where the man stands, a stream of four rods is going by him continually at lightning speed, about a mile a minute; hundreds of tons in twenty-four hours looping the loops through the rolls and finishing in red coils of quarter inch, lying innocent and rosy and round on the metal floor.

To the novice they look like wire; to the cognoscenti they are only rods, and in order to be wire some day are hustled off to the cleaning house and in bunches plunged into a bath of acid. This takes off the scale the rolling left on them. But acid in wire steel is like heresy in the church. It has to be purged away. This is done by immersion and then by a coating of lime to neutralize by chemical action whatever taint may remain. The steel is then baked from twenty-four to forty-eight hours to remove the hydrogen.

Wire making has just begun. From this time on it is a wonder-work to the novice, a mechanical sleight of hand performance by which hundreds of shadowy men and other hundreds of whirling wheels spin the rod down ever smaller and smaller till what was once a stodgy four foot billet is perhaps a thousandth of an inch thick, fifteen odd thousand miles long, weighs less than a quarter of an ounce to the mile, and has to be looked for with your best reading glasses. It is just three times as fine as the hair on your head.

THE WORK OF THE WIRE DOCTORS

Never think that the tall chimney of a manufacturing plant tells the story of all that goes on in its shadow. It isn't all coarse work. If you could see the things that are done to a block of steel, and the brains that are mixed with it, in the Roebling plant, before it comes out and goes on its way, they would make you take off your hat to a piece of wire for the rest of your natural life. But it isn't all, what happens to the outside. There are wire doctors who follow the changing symptoms of the metal through its many processes, with diagnostic eye as keen as any medico's for traces of typhoid or mumps. Through all the process there are reheatings and coolings, at carefully specified temperatures, to give temper and then to take it away, to keep the ductility without sacrificing endurance. It is one business where you simply have to eat the cake and keep it, too.

There is wet drawn wire and dry drawn wire, and chemical reasons for drawing wire wet, and divers ways of drying wet wire to attain certain conditions; there is lubrication by means of dry materials as well as oil, and soap suds, funny things that also act on the material itself in mysterious ways. But this is no text book.

TRAMWAY RUNNING ON WIRE ROPE CABLE DUMPING COAL AT MINE

No thinkable effort is omitted that will help to make the wire material perfect in quality and service condition, but the proof of the pudding in the making of wire is in the Olsen machines—miraculous things that will smash a big wire rope or snap a hair of wire and register to a decimal the breaking strength of each. There are tests for tensile strength, for torsion to show how many twists a piece of wire will stand, and for bending. There are microscopic tests for molecular condition and men who will almost tell you from a microscopic section the maximum service of which the rope made from a given wire is capable. Any bundle of wire that doesn't pass the test for the job on hand is discarded and used for something else, and a record of it all is kept with scrupulous care. Any foot of wire that passes through the shipping room on the way to market has a clean bill of health, ample for the use to which it is destined, and the amount of material that is scrapped for faults, where work is on stringent specifications, would be sudden death to a business that hadn't a wide range of uses for product of whatever quality. Fortunately for the users of high-grade wire the market for the lower grades is always hungry and crying for more.

THE WONDER OF DUCTILITY

There are complexities without end in the making and finishing of wire, but the real wonder of it lies after all in the initial principle which the German inventor in Bavaria gave to the world six hundred years ago—the simple but even now almost incredible fact that a rod of cold steel of the hardest quality—plow steel is the convincing name for it—can be seized by its sharpened end with a clamp they call a dog and drawn through a smaller hole, in a still harder piece of steel, three or four feet until it can be fastened to a drum, and then be wound off in miles almost without interruption. It is a wonder that grows as you watch it and yet it seems so simple. To see that steel, of tremendous strength and hardness, drawn through a tiny hole as if it were molasses candy—and yet it may have a tensile strength of two or three hundred thousand pounds to the square inch.

There is nothing spectacular about the wire mill where this is done. On long benches the die-holding appliances are set up and the dies set into them. The wire—or at first the rod—is run from a portable bobbin they call a swift, that stands on the floor, and the wire, after it has been given the hole, passes to a bobbin they call a block. Then it is taken on to a still smaller die and the same process repeated, with occasional reheatings, until it has the diameter of a thread.

CUTTING THE DIES

But by and by the time comes when the wire is so fine it cuts the steel of the die and loses its rotundity. Then a harder material is needed and the wire drawer goes the whole figure and uses a diamond. Cutting the steel dies is a cunning craft enough, but the expert, who, with a hair-like drill and a dab of diamond dust can penetrate a diamond with an opening that will be regular and measure to a thousandth of an inch, is a man who would think it no trick at all to pass a well fed camel through the needle's eye.

It would take a larger book than this to tell all the things that are done in the making of wire for various uses. In the main, the entire volume produced either goes to market as wire of one sort and another, to be applied to its various objects or for sale, or else it is twisted into rope, of which the Roebling Company manufactures four hundred kinds, sizes and many qualities. The common fence wires are not among the Roebling specialties, but wire nettings are manufactured from a soft variety of basic steel which lends itself to the weaving process with almost the ease of animal and vegetable fibres.

THE ENDLESS MANUFACTURES FROM "FLAT WIRE"

The "flat wire," which has now attained immense volume of production, is, for the most part, rolled down from the round, in many qualities, and shipped as material to the makers of many things. There are wide, thin, beautiful ribbons which find their way to the shoestring factories and are cut and clinched to the laces as tips. The list of novelties and parts that are made from various forms and widths of flat wire is as long as the list of Smiths in a New York directory. In the novelty shop, which does a million things, wires are cut and mechanically bent in hundreds of thousands of shapes, for clothes hangers, pail ear staples, daubers for bottles, meat skewers, hog rings, thread guards for textile machinery, basket fasteners, shackles for car seals, saddlery parts, Welsbach mantles, clips and links for bedsprings, wiring for toys of all descriptions—and so on and on and on. And all this novelty business is a side line, like the square and triangular wires that are used by oil well drillers to keep the sand from getting into the oil.

The special shapes of high quality wire that are made to order, to provide hard-wearing parts for typewriters and many other machines, are almost without number.

SALVAGING "MILL ENDS"

With the increasing cost of labor and materials effort has been made to salvage and make use of "mill ends" of wire, running sometimes to large quantity, which formerly were accounted waste. These are now passed through a straightening machine, which lays them out in uniform bundles of some ten feet in length, which again may be cut to shorter lengths for special purposes. In the buildings where this is done, at the Upper Plant, are piles of neat bundles of all shapes and sizes and grades, which once went to the scrap for reworking but now are utilized without additional cost.

COPPER WIRE AND COPPER ROPE

Copper wire is manufactured by the Roebling mills in very large quantities and in many sizes and forms, principally for electrical use and for service where water corrosion shortens the life of steel. The little bond wires that link the rails of railways to perfect the carriage of current in the block signal system are mostly steel, but copper is used at stations and on sidings where the leakage from standing cars is apt to contain acids. Copper wire of all sizes down to the very fine is spooled and sold for use in arts and manufactures. For marine uses a deal of copper rope is made, and copper strand is twisted for lightning rods, the fixtures and supports of which, in turn, are manufactured from round and flat steel wire. The piles of this equipment, waiting shipment in the Roebling storerooms, give proof that the satire of the cartoonist and the mockery of the funny writer cannot destroy an ancient faith.

The telephone and telegraph companies use uncountable miles of copper wire in line service and other miles in fine sizes for instrument coils and divers other functions. Electricity as an agent would be a halting cripple without wire. The dynamo would have little more utility than a washtub.

Armatures, frames for which are formed from flat steel wire, are wound in the Roebling plant in impressive number.

One of the largest fields for copper is trolley wires, which are of great size and of many eccentric shapes.

This is merely a glimpse at the utilities that go to make up the field for Roebling wire. It is doubtful if today the company owns a complete list of the wire it has made for special and even eccentric purposes, or knows within many thousands the things that are manufactured from its wire product after it leaves the shipping room.

COATING AND FINISHING

Use determines much in the finishing of wire, and of wire rope as well, as not alone concerning the chemistry of the inside, but the covering of the outside. Material that is made for service out of doors, under water or under ground, to ensure long life needs an exposed surface more resistant to moisture than the naked steel. Copper is proof, but the pure wire is expensive for most uses and where severe strains are incurred it lacks in strength. Modern science has been too busy to recover the art of hardening copper which the ancient Egyptians lost.

Zinc, in its best application, makes steel wire weatherproof for many years and the apparently simple process of galvanizing, the fixing of a coating of zinc on the steel has multiplied many fold the utility of steel wire in places where it could ill be spared. But there is galvanizing and "galvanizing." The first is worth the money it costs.

There are other coated wires, too. The aeroplane strands and cords are tinned. There is a bronze enamel, and a copper coating which looks as if it were applied for protection but is really the incidental result of a dip in sulphate of copper, for other purposes in the course of fabrication. The coating of wires is chiefly done in the wire works of the Kinkora Mills, though a galvanizing house is maintained also at the Upper Works. For wire that is to be made into galvanized ropes and cords, the galvanic treatment is given before it goes to be made up.

JOURNEYING THROUGH THE ROEBLING PLANTS

For exercise, a journey through any one of the Roebling plants, and especially the great Upper Works, is as good as thirty-six holes of golf. It is upstairs and downstairs, over an interminable number of thousands of square feet, through the mazes of a picture that is always changing its detail and its rate of speed, but which is all centered on one idea, to keep the stream of wire and wire rope, of all sizes, kinds and colors, moving toward the shipping room. It all seems so easy in its progress, so free from friction or any trace of confusion, that the layman does not stop to consider how many problems have bobbed up along the way of production, even of the most modest wires and rope. Wire is a trade involving intimate knowledge of many lines of business and manufacture, since the character of wire required differs in nearly all.

HOISTING FULLY COMPLETED LOCOMOTIVE WITH WIRE ROPE SLING

To the novice, wire is wire. Here he learns that what is wire for one thing is valueless for another and wire that looks to the unpracticed eye as if it were ready for market always has to undergo a few more processes before it is up to demands. Wherever, however far, you travel in this succession of high-roofed, airy buildings, you come always upon some new regiment of machines, some new container of chemical or metal, with a long line of reels unwinding wire to undergo some additional treatment. And always moving among the buildings are cars, big and little, packing wire or material from one place to another, to feed the wheels and furnaces. The tonnage from plant to plant and from house to house in the Roebling works would make a first-class annual business for many a modest railroad, even if it carried nothing else.

INSULATION

But when wire is finished it isn't always finished. Since electricity spread itself over the earth in a million services, insulation in various forms has come to be almost as important as the wire itself. Insulation in its more advanced forms is a complex affair, gauged to accord with specific conditions and multiplying processes to secure the maximum of protection, both from electric current to life and property and from dampness and abrasion to the wire itself. In the making of wire screens the wire men have taken a leaf from the cloth-mill book, but in weaving a casing of cotton or other fibre around the wire for insulation the process is strongly reminiscent of some of the New England textile mills. Long rows of machines, black and silent and swift, reaching upward toward the ceiling, revolving rapidly on an upright shaft; long arms trailing downward, with wheels and bobbins like fingers plying dizzily but swiftly in and out around the wire which unwinds from its spool and keeps forever climbing. It is all like a Maypole, and the bobbins go in and out like children carrying each its ribbon. As the wire climbs, the whirling fingers braid around it a coating, tight fitting and impervious. Sometimes, where double insulation is required, there are two sets of arms, one above the other, the upper one putting on a second covering outside the first, of cotton of one color or another, or hemp or whatever else the experimentalists have found best for the purpose. You wonder how the bellcord in the railroad train can ever stand the pulling and jerking and wear and tear it gets. It is simple. It is just a perfectly made and highly tinned wire rope, with a double coat of braided cotton over it. The jacket may wear off in time, but the Roebling rope inside will never fail in a lifetime to get the message to the engineer.

When these snug coverings are finished the wire for certain uses is taken to another part of the works where it is unwound once more to pass through a bath of asphalt compound. After this process, which leaves a dull, dirty-looking surface, the spools of treated wire are put aside for drying, and then a final surfacing applied. The next journey is to the packing room.

TELEPHONE CABLES

Insulation is a wide range business. It cases wire in asbestos to prevent fire from stopping its work; but perhaps the highest phase is reached in the great cables of copper wires used in telephone service. For these the individual wires are covered with paper of various colors, which serves not only for protection but enables men at the opposite ends of a long cable to pick out unerringly the wires with which connection is to be made. Colors are few but possible combinations are many. The machining of this is more than ever like the Maypole, with pink and blue and yellow strips of paper flashing in the shadows. When the wires, paper covered, are brought together in the cable, sometimes three or four hundred of them altogether, the whole goes through the taping machines, which apply one or two suits of what may be called "underwear," for after it has been covered with two or three different materials there remains a suit of lead to be fitted, and this is a big work done deftly.

Who has not seen men in the streets dragging huge pipes of lead through the open manholes from big wooden spools? These are the cables you talk over. They have been papered and clothed—and tarred and feathered, maybe—and then encased in lead by a process that is so easy as to be laughable, and yet as ingenious as any one thing the wire miller does. Unrolling slowly from its spool, the heavy cable moves up to a machine built strong and four-legged from the floor. In the mid height of this, a few feet above the floor, is a square chamber containing molten lead. The cable passes in at the rear and upward. It requires some credulity to believe that it is the movement of the molten lead that carries the cable along, but in any case when it emerges from the "box," through an aperture that trims the soft metal down to uniformity, it has a solid lead covering as even as lead pipe, and at the point of egress cold water playing from just above cools it. Then it passes on through a long tank of water for final hardening and is wound slowly, clean and shining, on the great spools that are to carry it to market.

Many astonishing things are done in wire works, but done so swiftly, and smoothly and in such volume that they look easy. The man in the street, hurrying about his own business, never even takes time to wonder to himself how they are accomplished.

CHAPTER V. WIRE ROPE—THE GIANT.

"Pig" and "ore" and melting materials, with a condiment of carbon, are the body and bones of steel wire. Their virtues, combined and intensified by tireless processes, and tested unsparingly at every stage, are united in wire rope; and wire rope, when all is said and done, is the mighty backbone of the wire industry.

Wire rope to the multitude is simply wire rope. But one rope is no more like another than Jones is like Brown or Smith like Robinson. Wire rope is a combination of twisted wires, just as men are bipeds. That is where the similarity ends. In outward appearance as well as inward character, habit, tendencies and behavior in emergencies, wire ropes differ as widely as do people, and each has a meaning of its own.

Each also is the fruit of long study and repeated test of the work it is to do not alone on machines and in the laboratory, but under actual conditions of operation. The wire rope engineer will tell you every rope has temperament. He spends his life knowing other people's business—rope business—and working out their rope problems. The answers to these problems are the four hundred different sizes and kinds of rope that the Roebling Company manufactures on its regular schedules. The rest are specials. Go where you will in the world nowadays, you will find wire rope doing the work.

WIRE ROPE PROBLEMS AND THE ENGINEERING DEPARTMENT

With the completion of the Williamsburg Bridge, the Roebling Company withdrew from competitive fields of contract engineering, but it maintains a large engineering department and is ceaselessly busy with construction and installation problems from all over the world. In its files there is exhaustive record of every contract of magnitude, for construction, haulage, mine work, ship work—for any sort of work where rope is used and where the problems are difficult. Roebling engineers are always on the go, studying conditions where rope is to be used, to prescribe the fabric that will meet the need.

THE AEROPLANE – A WIRE ROPE CREATION

There is, to begin with, a questionnaire of ninety-three questions, to be filled out by the master mechanic or engineer on any special work for which rope is to be recommended and manufactured. When these are answered the engineer is ready to begin work, which starts with the selection of materials and does not end till the man who is to use it has had specific instruction as to its peculiarities and care and protection.

For this service the Roebling Company maintains a large corps of specialized engineers busily engaged solving the problems of wire rope usage, and making suggestions to effect economies in wire rope operation.

In fact, it doesn't end there. It is a saying in the Roebling establishment that a rope is never sold until it's worn out.

THE "LAY" OF THE ROPE

The cut ends of a diversified lot of wire ropes resemble, more than anything else, the eccentric forms of snow flakes, in their regularity and the grouping of their parts around a center. But there is nothing haphazard about the formations. Even the core is figured in the number of days it will add to the rope's life under varying conditions. The wide difference in ropes consists not only in the materials employed, which have much to do with their resistance to divers strains and the manner of their use, but in skillful selection of sizes in the wire and arrangement in the strands of which they are composed; again in the distribution in the strands, the twists of the strands themselves and the "lay" or manner in which these are twisted to make the rope. It is all the result of careful calculation.

THE CORE

A paramount factor too is the core, in securing the maximum of wear. Its mission, in most ropes, is not to add strength, but pliability, and to serve as a cushion to absorb the impact which the strands make under the tension of service. The fibre cores, for this reason, are usually treated with some lubricant. In the majority of ropes hemp is used for a core but in those intended for stationary service the core may be of steel. This will add from seven to ten per cent to strength and very largely to rigidity.

When we speak of wire rope most of us have a mental picture of a round fabric, but there are flat ropes as well, for use in mines or quarries where the haul is from great depths and twisting is to be avoided. These are made in all widths and thicknesses, and are constructed by placing several strands together, side by side, and sewing them together with soft iron wire. But it is the round rope that supplies the great demand.

THE STRAND

In considering rope, one may start with the strand. Strands, as may be seen from the pictures of transverse sections of ropes, vary infinitely in character, but always with a purpose. They are made up in ordinary practice, of four, seven, twelve, nineteen or thirty-seven wires, according to the work the rope is meant to do. In the rope mills you come upon long, low "stranding machines," reaching down a long room and carrying in horizontal arrangement, wide apart but in circular formation, the wires that are to form the strand. At a point carefully determined with reference to the strain on each wire, in order to preserve uniformity, all these wires come together and pass through one opening in a twisting machine which whirls them into a unit. The finished strand is wound on bobbins.

The direction of the twist, whether to right or left, is of moment in determining the character of the finished product.

"STANDARD" OR GENERAL PURPOSE ROPE

"Standard rope," so called, the general purpose rope, is composed of six wire strands and a hemp core, all being practically of the same size; but to secure particular results the number of strands may be four, five, eight, twelve or whatever may be desired. Already it will be apparent that there is wide latitude in rope making for the exercise of skill and the utilization of experimental record. This freedom in selection and adjustment extends through almost every process. For example, in the twists: when wires in the strands and strands in the rope are twisted in the same direction, which ordinarily they are not, the rope has what is known as a "Lang lay," after a rope man who devised the system. The twist, whether in strand or rope, has distinct effect in service. It may be long or short. If it is long the rope will be stronger and more rigid, if short, it will gain in flexibility. When it comes to the short twist rope, one sees the particular value of the twisting tests which were applied and recorded away back in the wire stage.

TESTING THE ROPE

It is singular, but it is true, that the aggregate strength of all the wires that go to make up a rope cannot be retained in the rope, at least in the laboratory on the testing machines. When the rope is tested for breaking strength it is found that no sample will show more than ninety per cent of the total, and the average is about eighty-two. Part of this failure is due to the angle of wires in the strand, with a resultant stress on wires in excess of applied load; therefore, the greater the number of wires in the rope, the lower the efficiency. The other reason is that the contiguous strands in the rope nick each other under high tension, and so are weakened. This, however, may not be important in ordinary working loads under service conditions. These casual truths show with what multiplicity of tendencies the rope maker has to deal in devising a product to give service and safety in the often ticklish jobs it has to do, with great weights in hand, and human lives at stake.

THE STEAM SHOVEL SHOVELS BY MEANS OF A WIRE ROPE

From molecular condition, as revealed by the microscope, down to the last petty detail in the plan of construction, there is never an end to the problems, and gravity has to be figured into the lifetime of a rope as surely as the elusive trace of sulphuric and muriatic acid producing hydrogen occlusion. Wire rope is a business of exactitude and eternal vigilance. You have to deal with breaking strengths of from 40,000 to 340,000 pounds to the square inch of transverse section, but the wire that will lift weights at the rate of more than a hundred tons has entirely different characteristics than the lower strength material. And the why of that must be traced back to the treatment of the steel when it was passing through the wire stage.

Rope makers dealt with molecules once and thought they were taking pains. They found they had to go back to atoms to handle their problems. Today the secret seems to lurk in the electron.

FITTING THE ROPE TO ITS WORK

Of the tricks in making ropes, there is no end. They are fitted for their work like a soldier or a gymnast, and built for it. A tiller rope must be flexible to the last degree, but it must be strong enough so it will stand up under the swift tensions of a storm or in the lightning manœuvers of a race. Therefore, like a few ropes built for other purposes, the composite parts are not mere strands of wire, but little ropes in themselves, complete in all parts. And again, while ropes exposed to weather and stationary, like ships' standing riggings, are galvanized, those that are subjected to constant bending are not. For every variation, there's a reason.

To the average man or woman, the elevators in tall buildings suggest danger. The rope engineer counts them highly safe because each elevator is equipped with a multiplicity of ropes and safety devices. What taxes his conscience and spurs him to the last possible effort, is the rope that goes to the "deep shaft" service, where the lives of men going up and down in five thousand feet or more of subterranean darkness, hang on the accuracy of his calculation.

Only now, the Roebling engineers will tell you, is wire rope being perfected. Much of it is in what seem to be small details of construction, which nevertheless go down into the basic principles that make for efficiency. Rope making has been treated as an exact science, because it dealt with materials that were more or less standardized. They are learning now that rope has a large unknown quantity that defies formula past a certain point. For the lack of a better term, they call it "personality." The labor of today, and many years to come, is to identify these intangible factors and bring them where they can be computed to the end of securing greater endurance and safety.

In the Roebling shops there are men working who got their jobs almost by heredity. Their fathers and grandfathers worked for John A. Roebling.

"You ask them," said the Chief Engineer, "why they do a thing a certain way. They tell you simply that 'that's the way to do it.'" In the old days John A. Roebling figured out the way, and gave it to his workmen in the shape of orders—today somewhat different methods are utilized. To the cumulative experience of over eighty years of wire-rope making, the Roeblings have always availed themselves of the latest engineering skill. With up-to-date research, chemical and metallurgical laboratories, every progress in the art has been incorporated in their product.

FOLLOWING THE ROPE AND ITS USES

The Roebling people say that wire rope is their "baby." They give it the utmost of skill and care and caution in the making, and then to see that these are not wasted, they follow it into the field, where it is to serve, with personal attention to its installation and with the most detailed instruction for its protection and use, figuring out with nicety the speeds to be maintained, the size of the sheaves or drums around which it should travel to minimize the strain, prescribing its lubrication, providing printed warnings against all forms of misuse or neglect, with pictures to show the reason why, and other instructions and pictures to aid in detection of the first signs of trouble or exhaustion, and the reasons therefor. Study of the Roebling method, from the ore yard to the field of operation, makes clear the reason why Roebling rope, from the very beginning of the manufacture, has been accounted standard for quality.

A Roebling catalogue is never complete. It cannot list and illustrate, without competing in size with the unabridged, more than a small part of the uses for which rope—and much of it special rope—is made, or the infinite number of attachments and accessories provided for installation and use on the job.

WIRE ROPE AND ITS WORK

There is the transmission of power by means of a round, endless rope, running at high velocity over a series of sheaves or pulleys, carrying power to a distance of three miles; there is underground haulage, for which five distinct types of rope are used, enabling the engineer to make light of grades, even with staggering loads; logging, in which, in the primeval forests of the Northwest, the horse or ox is a pigmy, and where the giant trunks, seven, eight or nine feet in diameter, are whisked up at the sides of mountains, hoisted into the air and deposited on cars, to be run down to the rivers on steep inclines, again operated by rope of great size and strength. There is quarrying, where rope is used in quantity for guying, and for hoisting the blocks of stone out of their beds, and then on aerial cable ways, to carry them on high over long distances to be loaded; there are the oil fields, in which just now, in the mad search for petroleum to supply the world's shortage, interminable miles of wire rope are being used, some of it an inch thick or over, to carry the drills, or for casing and sand lines. There is shipping—the battleship and the merchantman and the liner; the yacht, the riverman and the tug—all strung with wire rope from stem to stern, and some of them from truck to keel as well—not to mention mooring lines which have their own plan and formula; there is towing, to which wire rope brought new possibilities and freedom from old troubles and old perils—witness the towing of the dry dock "Dewey" from Chesapeake Bay to the Philippines, thirteen thousand miles, on a pair of 1200 foot Roebling hawsers, which stuck to their jobs without interruption, through all sorts of weather, and lugged their burden into the harbor of Olongapo without a sign of weakness or exhaustion; there is dredging, for which wire rope has largely supplanted the old and cumbrous chain which was never any stronger than its weakest link. There is hardly an important harbor in the world today where these stout ropes are not busy clearing pathway and anchorage for marine commerce.

MORE USES OF WIRE ROPE

The list does not end. There are incline railways, in the mountains of East and West alike, as well as in foreign countries, which have made mountain climbing a primitive form of sport, and enabled one-legged men with perfect ease to get the view from towering peaks which otherwise would have been accessible only to the hardy mountain climber; there are cable railways with which engineers have been able to run cars out on an aerial roadbed of wire, over impassable gorges and morasses, to make fills for railway or other construction; cableways, the forms and uses of which, in transferring materials, are without number; tramways and traction systems, which have now, save in particular instances, given way to trolley, and the copper wire for this, again, comes in large and continuous tonnage from the Roebling mills; there is the perfect litter of hoisting slings, all over creation, for wherever men are doing work or business of any kind, there is a load to lift, and the wire rope, with its special appliances for quick hitch and release, is fast relegating the old time chain to the category of antiquities. In 1862 the first of elevator ropes was made. Today millions are in use.

It is a long story, and one variety of rope is never just like another, save for the general purpose product before referred to, which figures in the schedules as "Standard." But in the making of all the many hundred kinds, the process, to outward appearance, is the same, and impressive in the simplicity to which it has been reduced. From the tiny specimen, made for some finical scientific experiment, to the three-inch monster that contains single wires nearly a quarter of an inch in diameter, and drags half a million pounds of ore, with the aid of powerful machinery, at the Spanish American Iron Company's mines in Cuba, the general principle of manufacture and the mechanism used in the making are all alike.

ROPE-MAKING MACHINES

In the several rope mills of the Roebling works are a large number of machines, some of which, built by John A. Roebling in the early days of his rope making, are still turning out rope, and good rope. His first product was made by hand in the old "rope walk" way. Today the ground where he did it is covered with buildings full of speeding machinery that has little rest—devices that stand in long rows, eating up the strand that unwinds from the whirling bobbins to feed it, and turning off steadily the completed rope, which passes to spools, large or small, in proportion to its weight and size.

Simply described, the rope machine pictures itself as a hollow column cylinder, strongly framed and braced steel from the base of which arms extend, like the lower branches of a spruce tree. At the ends of these the bobbins are rigged, carrying the strands which are to be twisted into rope.

These are led from the bobbins in toward the center, and pass into the column, which carries also the core and which in its turning twists the strands together. The complete rope passes out over a pulley on to the spools. Machines for the smaller sizes of rope are strung out in a long file. The larger ones require elbow room; each of those for the making of the largest rope has a room to itself and is installed on a foundation of steel and concrete.

HOISTING A HUGE NAVAL GUN WITH WIRE ROPE SLING

When the mechanism is at work it suggests somehow the solar rotations. The bobbins have a triple motion. On the ends of the arms to which they are attached they travel around the column, at a rate of speed which of course is determined by the "lay" required, but they are unwinding as the strand pays out and also turn completely end for end, at predetermined intervals. In the more modern machines there are two sets of arms or "branches" above the first, for the purpose of carrying a greater number of strands. In this type the arms carrying the bobbins are somewhat shorter, allowing for a great rate of speed. There is something mysterious in the sight of these flying reels of steel, or copper maybe, for many ropes of substantial size are made of copper for marine use, whizzing round and round like indefatigable moths around a big steel candle, or a dervish round his own spinal column on a spot of ground the size of a dinner plate, and the rope, hard, shining, round, packed around its core of hemp or steel, noiselessly gathering all this strength and energy into itself for use in the days of need. When you see it on the spool at the side, shining with its coating of lubricant, ready for work and able to do it, it is a little hard to associate so respectable and dignified a fabric with the rusty heap of iron that lay in the Kinkora yard.

SPECIAL CONSTRUCTION

There are records in the Roebling offices that tell interesting tales of special constructions, and pictures of enormous spools of rope, thousands of feet, in big diameters, running from spool to spool and since one spool is an ample carload, from one flat car to another, when loaded for shipment.

Such were the huge street railway cables, made for Australia, for Kansas City, for Chicago and New York. There is an amusing story of the New York street railway engineer who insisted that the cable be made in one section, 33,000 feet in length, but who changed his mind about the beauty of it when he got the goods and saw the elephantine spools of packed metal caving in the manholes in the city streets on their way to the point of installation. A gigantic rope machine was built in the Roebling plant to twist this mammoth.

The cars that carry these heavy cables were made specially for the purpose. An ordinary car would crumble under the load, but the machine and the cars are still in use, and busy.

When cables for street railways were discarded in favor of trolley, wire rope men thought the day of doom had come, but the field for wire rope for other uses has widened so fast and so far, in a rapidly widening world, that the cable orders, big as they were, have never been missed. It furnishes a significant index of the growth in all industrial activity, for there is no new phase of development or manufacture or work of any kind in which wire rope, or wire in some form or other, does not play an indispensable part.

POWER IN THE ROEBLING PLANTS

In the three Roebling plants there are four electric power stations, aggregating over 16,000 horse-power, and more than 150 boilers with 25,000 horse-power. The coal consumption on the three plants is approximately 1000 tons a day, and the fuel oil consumption about 20,000,000 gallons per year. In the Kinkora plant at Roebling there are thirteen miles of standard gauge railroad track.

CHAPTER VI. WORKING FOR UNCLE SAM.

Of the load that war laid on productive industry, it is beyond question that wire, the country over, carried its share. In the retrospect, every man and every organization tries consciously or unconsciously to figure out what part individual effort contributed to the big result. Fortunately, perhaps, the question of relative accomplishment and of everybody's share in the outcome is one of the things that can never be settled, but in the picture war has left on the memory of those who lived it, wire and wire rope can never be very far away.

As wire pervades every industry of peace and every department of living, so in the headlong rush of war, whether by land or sea or in the air, it was the handy and dependable agent that made a thousand other things possible. Wire did its work not only up in the smoke and the agony of the western front, not only where the fleets battled against the lurking death, but along every line of plain toil by which the unhalting supply of materials, both for battle and sustenance, was kept flowing to the point of need.

When the big order for multiplication of output came, wire rope manufacturers were not told to make one thing and a lot of it, as so many industries were. The demands of war, on the contrary, added diversity to what was already one of the most diversified of products. Everything was special. Every day's new load was a brand new problem in manufacture and in construction as well—something that had not been produced before, or at very best a new adaptation which required special manufacture and new organization; this in a skilled industry, at a time when skilled labor of any kind was scarce.

A STORY THAT WILL NEVER BE TOLD

The story of this period will never be told in its entirety. The Army cannot tell it, nor the Navy. They never knew it. All they did was to call for the stuff and get it. The wire makers will never tell it because they are too busy supplying the demands of peace—the rebuilding of a wrecked world and the development of a new one. Already the picture, big and thrilling as it was, is growing dim, its detail disappearing in the hurry of industrial production, the solving of new problems, the supplying of demand. They look back over the old requisitions and specifications of the feverish days of 1917 and 1918 and are surprised to see that dust has gathered on them already; they count the figures of overwhelming volume which are their "war history" and wonder how in the world they ever did it.

THE DOUBLE BURDEN UPON WIRE'S BACK

What doubled the burden on wire's back was that every existing industry for which it had been making rope was "essential." The wire men looked around to find what they could cut out. There was nothing. To maintain the supply of oil, of coal, of ores, of food, to keep all kinds of transportation in full swing, to see that elevators kept running so that activity should not cease—these and a thousand other things were all essential to unity of effort and increase of production. Altogether the saving was trivial. They all had to be supplied, most of them double, and the Allies had been piling in orders. On top of this burly task came our own Government's great and variegated and undeniable demands for war supplies.

In the carrying of such a load the wire industry was hampered by the fewness of its plants and their distribution over the States, some of them far from points of ocean shipment. It was plain when America entered the war that only the most thorough co-ordination and centralized control of operation could make success possible; only the most economical arrangement of forces and distribution of materials.

The Iron and Steel Institute, at the request of the Government, formed a committee to manage the production and distribution of wire rope, and from the fifteenth of May, 1917, this committee had on its shoulders the making of wire necessary to keep the country's work going at full speed and to supply the needs for war, of whose extent or character nobody had any clear idea. Karl G. Roebling, of John A. Roebling's Sons Co., was made chairman of the committee.

THE WAR – ONE LONG COMMITTEE MEETING

Throughout the war, the Roebling offices in Trenton were headquarters for the entire business of wire rope supply. It was one long committee meeting, with production going on at utmost intensity all the time. Here came all the orders for wire rope from the several Government departments and the bureaus in those departments, each with its long array of specifications, all requiring shipment to divers points. Much of the work required also a great labor of cutting and attaching, and fittings by the hundreds of thousands, and all, without an exception that stands out in anybody's memory, wanted in the minimum of time.

It is the proud record of that committee that when the fighting ended in November of 1918 every order had been filled and delivery made on time. Industry has no story of accomplishment to tell that can be more creditable than this. The Roebling plants, near to the seaboard and equipped for specialization, were devoted almost wholly to the manufacture of war stuff, domestic industrial orders being transferred to inland factories.

THE RECORD OF PRODUCTION FOR WAR PURPOSES

The record of production, for war purposes alone, shows that the Roebling Company manufactured a very large percentage of the whole, which ran to unconscionable millions of feet. During the war the productive capacity of the plant was increased as much as seventy-five per cent, and the list of the employed at times ran close to ten thousand men. The numerical increase in men did not equal the growth in output. Here as well as in almost every line of industry the war furnished a revelation of the capacity of men for work. New lines of production, requiring skill, developed in common laborers, the only kind that at times could be obtained, a facility in production that before the pressure of war came to discover it would have been thought impossible.

In looking back over the war work it is plain that the service rendered by the company in manufacturing material for the Allies, prior to America's entrance into hostilities, was of large value in familiarizing it with forms of production afterward required for our own Army and Navy. Another thing which aided in meeting a vast demand was the unremitting attention which the company had given to the perfection of aircraft material, from the first successful flight of the Wright brothers at Kitty Hawk in 1903. At that time study and experiment had been started in the Roebling factories looking to the production of aircraft wire and strand and cord for all the different parts involved, which should combine the utmost strength with the minimum of weight, with special reference to the stresses peculiar to aviation work.

When the hour of need came, Roebling aircraft products had reached a stage of perfection which saved a world of hurried experimentation and development. It was a demonstration in preparedness, although up to 1914 the work had been done solely to keep industrial pace with a new and important development of mechanical science.

THE GOVERNMENT ESTABLISHMENTS THAT CALLED FOR WIRE AND ROPE

It is a fairly long list of Government establishments that is shown on the Roebling records as calling for war supply of wire rope. It includes, in the Navy Department, the Bureau of Steam Engineering, Bureau of Construction and Repair, Bureau of Ordnance, the Bureau of Yards and Docks and the Naval Aircraft Factory.

In the War Department were the following: Office of the Chief of Ordnance, Depot Quartermaster, Chief Signal Officer, Chief of Engineers, the Army Transport Service, the Quartermaster General's Office, the Signal Corps, the Aircraft Production Bureau, United States Engineer's Office, General Engineer Depot, Bureau of Insular Affairs, Procurement Division, the Balloon Department of the Aircraft Production Board, and the Director General of Military Railways.

Always there were the United States Shipping Board, the Emergency Fleet Corporation, and the demands for these alone were a business. In addition to all this the Committee made allocation of orders for the Argentine Naval Commission, the British War Commission, the Imperial Munitions Board, the Italian Commission and the Belgian Government.

It doesn't seem such a large roster, but it took a world of wire to go around it. A few figures out of the total allocations will suggest what the total demand was and the task that it involved.

SOME OF THE BIG DEMANDS

The first big call on the wire rope producers was for submarine nets to protect the fleet bases and harbors. There were supplied to the Navy, for this purpose, 2,820,520 feet of rope, and it was regular rope that was required in this service, for the German submarines had developed a way of slashing through the earlier and lighter nets. For the new type the rope ranged from an inch and a half to three-quarters of an inch; but it wasn't merely a matter of shipping reels of rope. Almost all of it had to be cut into lengths and attachments made, for these barriers were designed in sections. This necessitated, for the Navy order, 153,000 fittings. The Army Ordnance Bureau used nearly a million feet of rope for nettings, which was shipped to various coast forts. The whole volume of wire rope for nettings was furnished within four months.

CUP CHALLENGERS, DEFENDERS AND SAILING VESSELS OF ALL TYPES SECURE THEIR RIGGING WITH WIRE ROPE

Another interesting order was from the Quartermaster's Department, which called for 6,852,500 feet of rope and the manufacture of 300,000 pairs of traces, requiring 3,000,000 splices. These are what are called thimble splices, and, while fitting one of them is ordinarily half an hour's work, the Roebling plant, with a force chiefly of men who were utterly unskilled, was turning off ten thousand pairs of traces a day at the peak of production on this order. This harness, for artillery purposes, was on English designs, adopted after considerable delay, but by means of which a horse, when shot down, could be eliminated from the gun team in half a minute.

The Spruce Production Bureau took over 8,000,000 feet of rope, the Emergency Fleet more than 12,000,000 and the Fuel Administration drew on at the rate of 2,500 tons a month. And all the time the mines and mills and ordnance plants, locomotives, cranes and all other manufactures kept getting largely increased supplies of rope to carry on their own war-driven work. Altogether the orders come to a figure that is hard to visualize.

84,000,000 FEET OF ROPE AND A HALF MILLION FITTINGS

But the climax, the call that taxed the wire rope makers most heavily and kept the arc lights burning in the mills was for the 84,000,000 and odd feet of rope and half million fittings which were required by the Naval Establishment for the North Sea Mine Barrage, which put a prompt and distinguished shackle on the German submarines. The fitting of this rope was a task of moment, calling as it did for delivery of the rope in lengths and made up ready for attachment on the ingenious plan which the mine involved. It was all done with time to spare.

The Adriatic Barrage, an even more ambitious project since it dealt with a depth of 3,000 instead of 900 feet, was all ready to be laid when the Armistice was signed. This took over 12,000,000 feet of rope.

When the fighting stopped, there was a perfectly good mine barrage in the North Sea that had to be taken up and put out of commission. This called for 616,000 feet more of rope, with fittings to make it of use. Every mine was cancelled without a mishap, and there are now more than eighty million feet of "A No. 1" wire rope reposing at the bottom of the North Sea. But it did its work, capturing no less than seventeen German submarines in the first week.

AND MORE THAN WIRE ROPE WAS ASKED FOR

The Roebling plant, for the time, was given over to the manufacture of war necessities, hence its problems of material were made easy by the Director of Steel Supply. But the Roebling output for war purposes did not end with wire rope. In May, 1918, the company was employing close to ten thousand men, and in addition to rope making they were busy with the manufacture of immense quantities of steel strand, strand for outpost cables, copper strand, telephone wire, copper wire and miscellaneous wires of all descriptions, which were needed in the service at home and abroad.

A material part of the war work was the manufacture of wire especially for the field telegraph and telephone systems of the Signal Corps in Europe, where the American Army communications were the admiration of Europeans. This material possessed certain peculiar characteristics, and while speed in its production was an essential yet it was necessary that every strand be perfect, for the fate of armies rested upon it.

The manufacture of this wire involved a great deal of detail and intimate knowledge of all sorts of materials, for while copper is used for electrical transmission there is an exterior protection of other metals and materials, each of which has its peculiar manufacturing difficulties.

THE COMPOSITE STEEL AND COPPER STRAND

For example, the "Composite Steel and Copper Strand" wire used by the Army was made up as follows: There was a center wire of tinned copper with ten outside wires of tinned steel. This wire had a maximum weight of 75 pounds a mile with a maximum breaking weight of 300 pounds. Other types of wire were silk wrapped, covered with a rubber compound or with a covering of cotton braid treated with a waterproofing compound.

TO MEET THE SIGNAL CORPS' REQUIREMENTS

Take one type of the thousands manufactured by the Roebling Company and see what must be done to make the finished product for the Signal Corps. This process, which includes both the manufacture of steel wire for the outer protection and copper wire for transmission, may be divided into the following parts:

All steel materials are analyzed and inspected. Acid open hearth steel is made in ingot form in special furnaces. The steel is classified, and the ingots are reheated and rolled into billets, which are cropped to eliminate all segregation. The steel billets are reheated and rolled into rods of about 3/16 inch diameter. The rods are then tempered for wire drawing. Then comes an inspection and testing for physical characteristics of the metal, and the rods are cleaned in acid, washed, lime coated and left to dry.

These rods are then drawn cold through dies to intermediate sizes requiring a repetition of the tempering, inspecting and cleaning operations. There is another series of drawing and then the final one through the hardest and toughest dies obtainable to a diameter of 12/100 inch. At this diameter one foot of the original rod has been extended to about 350 feet.

Then comes another inspection and test of the mechanical properties. The wire is next cleaned in alkaline and acid solutions to remove all trace of the lubricants used in the wire drawing, and the wire is subjected to a bath in pure hot tin. Finally there is a Government inspection and test.

So much for the manufacture of steel wire. The copper first appears in bars, which are inspected and tested for their metallic purity. The bars are heated and rolled into rods of about 3/8 inch diameter. These rods are cleaned in acid baths to remove all scale, and the wire drawn with the necessary annealing and cleaning until wire that is only .0285 of an inch in diameter is the result.

The final drawing of this wire requires the use of diamond dies with the necessary equipment and great skill of the wire drawers in piercing these minute openings. The copper wire then is annealed free from all scale and discoloration, and the tin coat applied by means of a liquid tin bath. Then the Government inspectors test the copper wire.

Ten strands of the steel wire are twisted about the one copper wire, and the Government inspectors again make tests to see if the inner copper wire is intact and properly protected by steel wire. All grease is removed from the strand, and tussah silk wrapped over the whole. To this is applied a compound with 30 per cent rubber, which is later vulcanized. Then come inspections for mechanical injuries and electrical characteristics. The single conductors are braided, the braid waterproofed, polished, twined, inspected, reeled for shipment, inspected by the Government agents, packed, inspected again by the Government agents and finally shipped.

All this is done with a great deal of rapidity but with no less care, the skill obtained by the workmen only by years of experience and by the technical men only by years of study. It required a thorough knowledge of steel and of the materials entering into the manufacture of steel, such as ore, pig iron and fuel, as well as of the properties and tests and manufacture of copper, tin, rubber, cotton, and various lubricants. And in the more general use of wire and wire rope, a thoroughly comprehensive knowledge of many other materials, all mechanical and electrical phenomena in fact, are essential.

CHAPTER VII. A CITY BUILT OUT OF HAND.

All up and down the Delaware, between Trenton and Philadelphia, the "quality folks" in olden times used to build stately homes, with broad acres at their backs and looking lordly, with their Grecian porticos, out from the high banks that command the stream. You may see some of them yet, faded and old and full of family history, most of which was not so important as it seemed to the builders. In the little towns that you pass on the trolley and the Camden and Amboy road, there is a certain Eighteenth Century somnolence, and a dingy pride of priority. They sleep on, as if it were creditable not to be busy. Bordentown, a few minutes' ride from Trenton, sits complacent amid its memories of the Bonapartes. It is there you change for Roebling.

ROEBLING, THE TOWN, A STORY IN ITSELF

Roebling—the town, not the plant—to which some attention has been given, is a story in itself. It is an industrial disturbance in the quietude of a sleepy and beauteous country. It is a rattler of the dry bones of tradition, and pretty nearly the last word in corporation communities. Roebling maintains no staff of highbrow sociologists to discuss the things capital should do in order to make labor's pathway broad and bright. There's a town superintendent to look after things and he earns his pay.

BUILT TO MAKE WIRE AND ROPE

The town of Roebling was built to help along the making of wire and the wire rope. Making good rope, it is a good town, without any fanciful notions about "welfare work." The Delaware, flowing by in its beauty, accounts for part of this. But to the Roeblings the Delaware means plentiful water supply and river transportation. To the workmen in the big mills which lie just at the back of the town, and to their families, which grow phenomenally, it means bathing, boating, a cool breeze on stifling midsummer nights, and a panorama that never ceases to be lovely.

In both the city plants, as business grew, building followed building. A compact and populous section had grown up at Trenton. More buildings could not be crowded into the original ground space. More land was needed, and as usual in such cases, men with land to sell all along to the south of the Upper Works, saw the company's need and had a brain storm about what the footage was worth.

The Roeblings tried a little farther down stream. But down stream didn't mean down price. So they made a clean job of it. Ten miles down the river was a little old station called Kinkora, where the real estate infection had not appeared. There was land well up above high water, and plenty of it. The Delaware was very cheap down there, as compared with Trenton city water rates, to a concern that used as much water as all the rest of the city put together.

A LIKELY PLACE FOR A WIRE MILL

It was a likely place for a wire mill, but if a dozen strangers had struck Kinkora on the same evening the town would have had trouble to find beds for them all. It meant twenty miles rail travel a day for the workmen to live in Trenton. So the Roeblings decided to build. Charles G. Roebling was then alive. The new site and the planning and building of the town were his charge. But, again, they didn't go looking for any welfare engineers. The whole job of planning plant and town alike was done in the long engineering room of the Roebling offices. At first they called the plant the Kinkora. They do yet, off and on, but the mills were a little below the station, and when the new venture was well under way, and the machinery had begun to squeeze out wire, and perhaps a hundred brick houses of various types had been erected, the place had to have a station of its own. The Pennsylvania Railroad said it was Roebling, and stamped the tickets that way. Kinkora is wearing off. It is still a sleepy little station just up the line. Between it and Roebling there are a mile or so of distance and a whole century of time.

The name "Kinkora" harks back to the year 1000, when King "Brian Boru" of Ireland lost his life at the battle of Clontarf. His palace was named "Kinkora." In 1836 an ambitious Irishman named Rockefeller (not John D.) conceived the idea of an air line railroad from this spot where Roebling now stands to Atlantic City. In fond remembrance of Erin's Isle he named the terminus on the Delaware "Kinkora."

The enterprise itself died an early death.

The Roebling Company has more than 200 acres of land in the new settlement, enough, in all conscience, to accommodate as big a business as almost anyone would want to do, and houses to shelter all its workmen. If the company should ever find it good business to shake the dust of Trenton from its shoes altogether it certainly has a place to go.

NO TIME TO LET THE GRASS GROW

From the day when the thing was decided on, no grass grew under anybody's feet. There was sand along that bucolic and undeveloped river bank, sand that ran well back, getting more and more like loam as you left the river. It was broken and uneven. The freshets of centuries had left hollows here and hummocks there. They were levelled. The knolls—dunes they would call them along Lake Michigan—were scraped down and dumped into the swales, and the excess was thrown into a sedgy morass along the river front, to make it into solid ground and give a clean, healthy shore, which is now one of the chief charms of the place. For the sections where grass was meant to grow—for dooryards and the like—tons upon tons of "top soil" were brought in to give a fertile surface.

The mill buildings went up first, on a broad space of one hundred acres levelled off for them, and then the town began to grow. That was sixteen years ago, and it has kept on growing. Every year sees a lot of new houses, of various values, and one and all well built and comely. And in all grades they are better houses than a workman, or a mill boss either, can get anywhere else in America for the same money.

TO MAKE A PROFIT BUT TO SHOW A SAVING

That has been the doctrine from the beginning. Charles G. Roebling said at the time something to the effect that every workingman was a free moral agent, and didn't want to be tied to anybody's apron-strings, that he wanted a square deal and a chance to live his own life out of business hours, and to get the worth of his money when he spent it. "We purpose," he said, "to make a fair profit on our investment, but we can do that and still show a man a saving. And we stop there."

It doesn't take long to realize that the Roeblings are living up to the original schedule. The rents, the figures on all sorts of commodities at the "village store," which sells everything from a pork chop to a piano, and the drug store, which is just as "Riker-Hegeman" as any live town could wish, are all below the current price scale in the rest of the country, by a margin sufficient to mean something to a family when they "tote up" at the year's end.

Electric light, coal and the other things a man has to pay for in any town are charged for here, but it doesn't take a legislative fight or a big row in the newspapers to keep the price down where a man can afford to pay it. Water is supplied free. The idea is that the man owes the company nothing but good work in return for his pay. After quitting time he's his own boss. The company tries to make life in the town pleasant enough so that he'll be glad to live there, and think he has a good job. And it recognizes that life has many sides.

AND THE TOWN HAD A BAR

It was in pursuance of the general thesis that when the town opened it had a hotel with a bar. "There's no use," they said, "in trying to make a mollycoddle out of a mill man. When he wants a drink he's going to get it, especially the foreign born. We don't propose to pick his drinks for him. If he wants whiskey it's a good sight better for us that he should be able to get it here like a human being than to trail into Trenton and take a chance with the stuff that goes over the bars where a workingman drinks. The whiskey here isn't gilt-edged, but it's decent, and it's worth what it costs."

Prohibition settled the drink question, but while the cafe lasted in Roebling it kept the men from going to town to battle with the "embalming fluid," and not showing up for the customary three days. That too was good business.

FIRE, POLICE, BANKS, STREETS

After the dirt and noise and disorder of a city street, it is like a sedative to slip from the train into the peace and the wide spaces of Roebling. The tidy station is at one side, at the other, beyond the switch tracks, the little gate-house which gives ingress to the mill enclosure—if you have the proper kind of pass. From here a trim concrete walk leads on past the ground of the plant and its fence of tall pickets, toward the river, and the town. As you go, you meet with courtesy. It is not drawing the long bow to say everybody in Roebling—outwardly at least—is civil and good natured. Just beyond the mill grounds you come upon the police office, with trig coppers who seem to have very little to do. Like the shining fire engines, which stand in the adjoining building ready for service either in town or plant, they seem to be maintained chiefly for insurance and ornament. But they are practical organizations at that. The Roebling Company learned what fire was during the war, when two of the biggest buildings in the Upper Works were destroyed.

From this point the streets lead away, broad, clean streets with the best of sidewalks, and drainage. The town has spread out now so that it looks no more like a toy city. The streets are 80 feet wide, with the exception of Main Street and Fifth Avenue, which are 100 feet wide. Trees have been planted which already make it attractive. In front of every house is a dooryard, a patch of green grass to remind a man that God made the world.

HOUSES

Adjoining fire and police houses, there was formerly a trim little bank whose business has expanded to such an extent that it has been enabled to move to the centre of the business section of the town in an attractive and up-to-date building of its own.

The houses, while of widely different types, are for the most part made of brick. In order to avoid fire danger, the minimum of wood is used in all the buildings of the town. The houses are all constructed on the most improved plan of sanitation and hygiene. Through the block, giving access to the back-doors, run clean alleys, wide enough to allow wagons to pass for the delivery of coal, foodstuffs and other commodities, and for the collection of waste. The company is now halting between the erection of an incinerator plant to consume the garbage for its 700 and odd homes, or a "hog farm" as part of its three or four hundred acres, which without difficulty could turn out 1,000 to 2,000 head of swine a year, and further reduce the cost of living. It is possible, too, that it may some day produce its own milk.

There is a marked difference between some of the houses first erected and those of more recent construction. At present the "bungalow" type is in great favor, since it facilitates the labor of housekeeping. More pretentious dwellings, for the men holding important positions in the plant, are sufficient to make a rent-ridden, janitor-jaded, bell-boy bossed New Yorker wonder what he is being punished for. One handsome colonial home just built for a superintendent in one of the wire mills would be a credit to any commuter town.

BASEBALL, RECREATION BUILDING, THEATRE, BALLROOM

Always as you pass through airy Roebling you encounter some new institution built to make it seem like a regular place. There is a baseball ground which would be a credit to any city, with its tidy green grandstand and its carefully manicured diamond. The Wire Works team is now prominent in one of the State Leagues. There is a recreation building, with billiard and pool tables and the best bowling alleys that can be built. There is a spacious assembly hall, with theatre stage and a scrumptious curtain bearing a picture of the Roebling Brooklyn Bridge. The gallery is commodious. The seats are removable, leaving a ballroom of impressive size, and adjoining rooms are equipped with ranges, refrigerators and dishes for the preparation and service of suppers or of dinners great and small.

Take notice of the hotel, the boarding houses where single men live well and cheaply, of the public school, the hospital, the doctors, the nurses, the dispensary. And these last are busy functionaries.

VERY LITTLE SICKNESS, VERY MANY BABIES

There is very little sickness in Roebling. The sanitation is studiously good, but when you are sick there they look after you, which is also "good business," and babies are a favorite form of diversion. This is impressively true. You sense it wherever you go. There are children everywhere – good looking wholesome "kids." And something makes them glad to live here, too.

BEING A BOY SCOUT AT ROEBLING

To be a boy scout in Roebling is about as good fun as a boy could have. For a long time the company gave the boys too much. Then it woke up to the fact that half the sport of being a boy scout was to do things. So the Scouts were told if they wanted to keep the perfectly corking club house on the river bank, with its big meeting room, its open mouthed fireplace, its mounted deer heads, and banners, and books and guns and spears and swords and all the other junk the boy soul loves, they'd have to work for it. Goodness knows they do. The grounds around that shack in spring are turned up like a golf links. What they have done in the way of white birch rustic railings along the winding walks that lead to the grounds would make a Chippewa Indian sick with envy. This year they are to help build a long float from the club house to the water, to launch their canoes on.

To the medical equipment is added a hospital for contagious diseases, standing away out in the fields. And in the outskirts also is land set apart for gardens, where the millworkers have allotted plots of ground for the raising of their own vegetables. The manure from the stables, where sixty horses are kept, helps to make gardening worth while. Even to be a mule in Roebling is comfortable. There are old mules there – you see them just wandering around the paddocks, eating and growing older – that will never see thirty-five or forty again. Nobody ever will send them down the long trail. They have worked hard for the Roebling Company. It will feed them till they simply lie down and die of their own accord.

Feeding – whether mules or people – is habitual. When John A. Roebling first made rope, he had three or four men working with him. They had a table in the shop. As the business has grown, this custom has continued. Today the entire office force at the headquarters in Trenton—some 230 persons of all ranks—gets a dinner every day that for sheer quality cannot be equalled in any of the city hotels. It may be a fad to feed that whole crowd fresh yellow cream brought in every morning from the Roebling farms, but—it's good business.

THE PARK

The high land on the bluff overlooking the river at Roebling is a park, with trees and benches, and a place where the band can play while the folks sit taking the air on a hot summer night. In a neat enclosure of Roebling wire, convenient to all parts of the town, are tennis courts, for general use. There is a sanitary barber shop, where five shining chairs are always full. Roebling has the best barbered lot of foreign-born workmen in America.

HOW THE FOREIGNER LIVES IN ROEBLING

In a town like this are lessons for those who like to try to translate the foreigner for the good of American industry. There are those who cherish a superstition that the foreign workman in the United States lives poorly. In Roebling it is remarked that it is the foreigner who is the best customer in groceries and butcher's meat. He buys chickens instead of beef brisket, and not one chicken, but two and three. It is he also who buys the Hood River apples and the best grape fruit.

And as for bread—you should see the bakery. "Sunny Jim" would sing to see it—clean and shining, and turning out all kinds of bakestuffs besides the big round red-blond loaves of "European bread," which they say "has the strength" in it. The baker's wagon, loaded to the very top of the canvas cover, goes through the town and the workers' little children run homeward from it with two, three, four loaves altogether as big as themselves. Crescent rolls, which cost a nickel at a French bakery in New York, are sold here for two cents apiece.

So it goes in Roebling. Over on the one side are the negro quarters. They have everything anybody else has including a recreation house—and when they recreate, they just recreate.

If Roebling was an experiment, it is not so any longer. It is full of comfortable people, and in seventy years the Roebling theory as to what a workman wants and how he should be treated has never proved itself more conclusively than here. It is a suggestive fact that in all that time, save for some insignificant incidents, the Roeblings have been free from the nightmare of "labor troubles." It may be because its workmen have nothing worth while to complain of. Every effort is made to make them comfortable without making them feel like dependents.

It is the outworking of a great business theory. In these times it is of impressive significance.